CIPROFLOXACIN USERS GUIDEBOOK

Everything You Should Know About the Benefits, Safe Usage of Ciprofloxacin in Treating Bacterial Infections, Skin and bacterial infections, Respiratory Tract Infections, Urinary Tract & Sexually Transmitted Infections (STIs)

Dwight Manfredi

Copyright@ 2024

All rights reserved. except for a limited number of quotes used in reviews and a few other nonprofit uses permitted by copyright regulations, no part of this work may be reproduced, distributed, or transmitted in any way without prior written permission from the author.

The information of this book is intended solely for general education use. It is not meant to replace professional medical guidance, diagnosis, or treatment. If you have any queries concerning a medical matter, you should always speak with your health care practitioner.

Table of Contents

Copyright@ 2024 1
Chapter One 4
Introduction..4
    How Does Ciprofloxacin Work? 4
    Why is Ciprofloxacin Prescribed? 4
    Understanding Antibiotics 6
Chapter Two...7
Uses of Ciprofloxacin 7
    Urinary Tract Infections (UTIs)...........7
    Respiratory Tract Infections 9
    Gastrointestinal Infections 10
    Skin and Soft Tissue Infections 11
    Sexually Transmitted Infections (STIs)12
    Bone and Joint Infections 12
    Inhalational Anthrax 13
    Prostatitis 14
Chapter Three 16
Administration & Dosage 16
Chapter Four...20
Side Effects 20
    Minor Side Effects 20

Major Side Effects 21
Chapter Five **23**
Precautions 23
Chapter Six **28**
Ciprofloxacin Interaction 28
    Medications and Ciprofloxacin 28
    Supplements and Ciprofloxacin 32
    Food and Ciprofloxacin Interactions . 32
    Alcohol and Ciprofloxacin 34
    Herbal Supplements and Ciprofloxacin 34
    Prevention 36
Chapter Seven **36**
Overdose of Ciprofloxacin 37
    Symptoms 37
    What to Do in Case of an Overdose .. 39
    Treatment 39
    Prevention 41
    Potential Long-Term Effects of Overdose 43
Chapter Eight **44**
Storage 45
The End **53**

Chapter One

Introduction

Ciprofloxacin belongs to the broad-spectrum antibiotic class known as fluoroquinolones. In therapeutic settings, it is commonly used to treat a variety of bacterial infections.

Think of it as an effective means designed to find and destroy harmful bacteria in the body. By stopping bacteria from multiplying, it helps your body fight off infections that could otherwise leave you feeling weak, worn out, or uncomfortable.

How Does Ciprofloxacin Work?

Ciprofloxacin inhibits bacterial DNA gyrase and topoisomerase IV, which are essential for DNA replication, transcription, repair, and recombination in bacterial cells. By interfering with these enzymes, Ciprofloxacin inhibits bacterial cell division and ultimately results in cell death.

Given its bactericidal activity, which makes it effective against both Gram-negative and some Gram-positive bacteria, it has a wide range of clinical applications.

But when it comes to Ciprofloxacin, there isn't a "one-size-fits-all" strategy. This type of antibiotic is used by doctors to treat certain bacterial illnesses.

Why is Ciprofloxacin Prescribed?

Your doctor may prescribe Ciprofloxacin for a number of infections that might disrupt daily life, including skin infections, gastrointestinal issues, lung infections, urinary tract infections (UTIs), and other bacterial infections. Its ability to fight persistent microbes makes it an essential drug in contemporary medicine.

Understanding Antibiotics

prior to understanding more about Ciprofloxacin, it is also helpful to have a basic understanding of antibiotics in general. Antibiotics are medications that either kill or stop bacteria from growing.

Every antibiotic targets a distinct kind of bacteria and has benefits as well as downsides. Ciprofloxacin's strength is its ability to combat bacteria that cause serious illnesses, but proper administration is necessary to ensure that it works as intended without causing unwanted side effects.

Chapter Two

Uses of Ciprofloxacin

Making informed decisions about your health can be aided by being aware of the uses of Ciprofloxacin, a strong antibiotic used to treat a variety of bacterial infections.

Due to the fact that it treats a wide range of infections, it is a valuable aid in modern medicine. However, like with any medication, it is important to take it as directed and to be aware of its benefits and drawbacks.

Urinary Tract Infections (UTIs)

Ciprofloxacin is commonly used to treat both basic and complex urinary tract infections, including Pyelonephritis. UTIs can cause a great deal of discomfort, such as pain when peeing, frequent urges to urinate, and sometimes fever or back pain.

This antibiotic targets and eliminates the germs that cause these ailments once it reaches therapeutic concentrations in the urinary system.

For infections caused by susceptible Gram-negative uropathogens such as Escherichia coli and Proteus mirabilis, it is often advised.

For complicated situations, such as catheter-associated UTIs, Ciprofloxacin is still the recommended course of treatment, particularly for patients who are resistant to other antibiotics.

It's important to understand, though, that Ciprofloxacin should only be taken as prescribed by a doctor in order to treat a UTI, as abuse could lead to resistance or treatment failure.

Respiratory Tract Infections

Ciprofloxacin is an effective treatment for lower respiratory tract infections caused by susceptible Gram-negative bacteria, including Hemophilus influenzae and Klebsiella pneumoniae.

It is commonly used to treat chronic bronchitis exacerbations, community-acquired pneumonia (in certain cases), and bronchiectasis with bacterial infections on top of it. These conditions may have a major effect on your breathing, leaving you feeling worn out and tired.

Ciprofloxacin can aid in your recovery and restore your respiratory health by combating these diseases. Since not all respiratory infections are brought on by bacteria, it's imperative to make sure that Ciprofloxacin is the best choice for your specific illness.

However, microbial culture and sensitivity results must be used to guide its use in respiratory tract infections because it is less effective against Streptococcus pneumoniae than other respiratory fluoroquinolones.

Gastrointestinal Infections

The drug Cipro is used to treat gastrointestinal disorders, particularly infectious diarrhea caused by enteric bacteria such as Shigella dysenteriae, Salmonella infection, and Campylobacter jejuni.

It is also effective in treating traveler's diarrhea, particularly diarrhea caused by enterotoxigenic Escherichia coli.

Additionally, Ciprofloxacin may be advised for typhoid fever (enteric fever) caused by Salmonella typhoid or Salmonella paratyphoid in endemic areas.

While the antibiotic may help resolve the infection, it is important to stay hydrated and get adequate sleep throughout treatment.

Always follow your doctor's advice as your body heals from the internal strife that gastrointestinal disorders can cause.

Skin and Soft Tissue Infections

The skin is the biggest organ in the body, and cuts, scrapes, and other skin breaks can allow bacterial infections to spread.

Sometimes, when susceptible Gram-negative bacteria are the root cause, Ciprofloxacin is administered to treat infections of the skin and soft tissues, including cellulitis, abscesses, and wound infections.

These infections have the potential to spread and create serious issues if left untreated. Ciprofloxacin helps protect the skin and prevent the spread of diseases by fighting the causative bacteria.

It is frequently only administered to patients whose infections are caused by known Ciprofloxacin-susceptible organisms or in circumstances where other antibiotics are inappropriate.

Its ability to penetrate tissues makes it an excellent option for treating these types of infections.

Sexually Transmitted Infections (STIs)

Ciprofloxacin is sometimes used to treat certain STD's, including uncomplicated gonorrhea. A person's physical and emotional health can be greatly impacted by STIs.

Although current guidelines recommend utilizing alternative medications to treat gonorrhea, Ciprofloxacin may still be considered in specific circumstances if susceptibility is established.

However, the emergence of resistant strains of Neisseria gonorrhoea has significantly reduced its utility in this context.

Bone and Joint Infections

Although it is used less commonly, ciprofloxacin is used to treat infections of the bones and joints, such as osteomyelitis and septic arthritis. These painful and incapacitating infections may have a detrimental effect on mobility and quality of life.

It is commonly used in concert with other therapies to ensure comprehensive coverage, particularly in cases of chronic osteomyelitis where continued care is required.

Considering its oral bioavailability, ciprofloxacin is a useful option for patients transitioning from intravenous to oral therapy, allowing for the outpatient treatment of certain conditions.

Inhalational Anthrax

It has been approved to treat Inhalational anthrax (Bacillus anthracis) and, in very extreme cases, as part of post-exposure prophylaxis.

Given that anthrax is a potentially lethal disease, Ciprofloxacin's capacity to prevent its spread can be crucial in high-risk situations.

Treatment usually consists of high-dose Ciprofloxacin administered over a long period of time, often in combination with other antibiotics, to ensure the pathogen is eliminated.

Prostatitis

Ciprofloxacin is an excellent treatment for both acute and chronic bacterial prostatitis, particularly when the underlying cause is Escherichia coli and other sensitive Gram-negative bacteria.

Chronic prostatitis often requires long-term treatment to achieve adequate tissue penetration and bacterial elimination.

Ciprofloxacin is suitable for this indication due to its properties and is present in the prostate tissue in significant quantities.

Chapter Three

Administration & Dosage

Despite being a potent antibiotic, the effectiveness of Ciprofloxacin depends on how it is taken and dosed. Proper usage of this medication can be the difference between a long-term illness and a speedy recovery.

Ciprofloxacin comes in a variety of forms, each designed to satisfy particular needs and preferences. The suggested form will depend on the type and severity of the infection being treated. The most common kinds are as follows:

Tablets: Usually used to treat infections like skin, respiratory, and urinary tract infections, tablets are one of the most widely used forms of Ciprofloxacin. The tablets are easy to take and don't significantly disrupt daily routines.

Oral Suspension: For those who have difficulty swallowing tablets, Ciprofloxacin is also offered as a liquid oral suspension. This is particularly helpful for children or those with certain medical conditions that make taking medications difficult.

Administration via Intravenous (IV): In more serious situations, Ciprofloxacin may be administered intravenously (through an IV) in a hospital setting. This method delivers the medication directly into the bloodstream, ensuring prompt treatment against serious infections.

Ophthalmic (Eye Drops) and Otic (Ear Drops): To treat infections that impact these areas, Ciprofloxacin is also available as otic (ear drops) and ophthalmic (eye drops) drops. By specifically addressing bacterial infections in the eyes or ears, these medicines offer treatment at the source.

The amount of Ciprofloxacin you need to take depends on your age, weight, the kind and severity of your sickness, and any underlying medical conditions.

Your health care provider will carefully select the appropriate dosage to ensure effectiveness while minimizing side effects. Following their instructions to the letter is essential.

Adult dosages of oral Ciprofloxacin usually range from 250 mg to 750 mg twice a day. The typical dose for intravenous administration is 200–400 mg every 8–12 hours, depending on the severity of the infection.

Ciprofloxacin is generally not recommended for children unless there are certain situations when the benefits outweigh the risks, such as treating severe UTIs or anthrax exposure.

Pediatric dosages are calculated based on the child's weight and the type of infection being treated.

The length of time you need to take Ciprofloxacin may vary depending on the infection. For example, a simple UTI may require therapy for three to seven days, but more serious infections, such bone infections, may require weeks of care.

For people with renal impairment, dosage adjustments are necessary to prevent medication buildup, which may increase the risk of adverse effects.

Always complete the entire course as prescribed, even if you start to feel better before the medication is finished. Recurrence or antibiotic resistance may result from early withdrawal.

Chapter Four

Side Effects

Some people may have mild to severe side effects even though Ciprofloxacin is a successful treatment for a number of bacterial diseases.

Understanding these side effects, recognizing their symptoms, and knowing when to seek medical attention are crucial for the safe use of this antibiotic.

Despite being used to treat bacterial infections, Ciprofloxacin can interact with healthy cells and tissues, potentially leading to unexpected side effects.

Individual differences may exist in the way that Ciprofloxacin affects each person. Some persons experience side effects, while others do not, as a result of this variation.

Minor Side Effects

Although the most common side effects of Ciprofloxacin are usually mild and may resolve on their own without medical intervention, it is nonetheless important to monitor your health while taking the drug.

If you encounter any of the following, let your doctor know, but keep taking the drug as directed. These include:

- Nausea and Vomiting
- Diarrhea
- Headache
- Dizziness or Light-Headedness
- Restlessness or Insomnia
- Skin rash or Itching
- Abdominal pain

Although these side effects are often mild and temporary, it is important to notify your physician if they worsen or persist.

Major Side Effects

Ciprofloxacin seldom causes serious side effects, though some people may have them. By identifying these symptoms early on, more serious problems can be avoided.

Seek immediate medical attention if you experience any of the following:

- Tendon Rupture or Tendinitis
- Nerve Damage (Peripheral Neuropathy)
- Severe Diarrhea (C. difficile infection)
- Mental Health and Neurological Effects
- Severe Allergic Reaction (Anaphylaxis)

Serious side effects may require stopping the medicine and seeking immediate medical attention. It is critical that patients report any adverse reactions to ensure timely intervention and minimize complications.

Chapter Five

Precautions

Ciprofloxacin requires more than just taking it as directed by a doctor. You need to be aware of specific precautions in order to use this medication effectively and safely.

By understanding the precautions involved with using Ciprofloxacin, you may protect yourself from potential risks and ensure the safety and efficacy of your therapy.

This medication should not be taken by patients who have a history of known hypersensitivity to Ciprofloxacin, other quinolones, or any of the excipients in the formulation.

Patients with a history of severe hypersensitivity reactions, such as anaphylaxis, should not get Ciprofloxacin because these reactions have been reported.

Monitor Your Tendons

Particularly in patients undergoing corticosteroid therapy, elderly individuals (over 60), and those with a history of tendon disorders, Ciprofloxacin and other fluoroquinolones are associated with an increased risk of tendonitis and tendon rupture.

It should be advised that patients stop using Ciprofloxacin and see a doctor if they experience inflammation, edema, or tendon pain.

Central Nervous System

The central nervous system (CNS) adverse effects of Ciprofloxacin include headaches, tremors, dizziness, and, in rare cases, seizures or psychiatric symptoms such as confusion and hallucinations.

Patients with a history of disorders affecting the central nervous system, such as epilepsy, should be closely watched while taking Ciprofloxacin.

Peripheral neuropathy, characterized by pain, burning, tingling, numbness, or weakness, has been observed in patients using Ciprofloxacin.

If treatment is not initiated immediately, symptoms could develop soon after and become irreversible. If signs of neuropathy develop, Ciprofloxacin should be stopped to prevent irreparable damage.

Renal/ Kidney Functions

When using Ciprofloxacin, patients with liver impairment should exercise caution. Liver function tests should be performed on a regular basis, especially for patients who already have hepatic problems.

Since Ciprofloxacin is mostly removed via the kidneys, dosage adjustments are necessary for people with renal impairment in order to avoid drug accumulation and toxicity. Renal function monitoring is recommended during the course of treatment.

Pregnant and Breastfeeding Mothers

Ciprofloxacin is generally not recommended for use during pregnancy or breastfeeding since it can pass through the placenta or enter breast milk and may affect the development of the fetus or infant.

If you are pregnant, wish to become pregnant, or are nursing a newborn, make sure to inform your doctor before starting Ciprofloxacin.

Alcohol Consumption

When taking Ciprofloxacin, it is recommended to limit alcohol consumption. Some of the side effects of the medicine, such as headaches, nausea, and dizziness, can worsen if you drink alcohol.

Additionally, drinking too much alcohol might strain your liver and kidneys, which are already digesting the antibiotic. By limiting your alcohol intake, you can reduce the likelihood of problems and promote a smoother recovery.

Stay Hydrated

Ciprofloxacin may affect your kidneys, especially if you are dehydrated. When taking this medication, it's important to stay well hydrated to help your kidneys process and eliminate the medication.

Avoid dehydration-causing substances like alcohol and caffeinated beverages, and drink lots of water throughout the day. Maintaining adequate hydration can also help prevent negative side effects like nausea or dizziness.

Chapter Six

Ciprofloxacin Interaction

Understanding how any prescription interacts with other drugs, foods, or supplements is crucial for your health and safety.

Ciprofloxacin is a powerful and effective antibiotic, but it's not an exception. These interactions can be dangerous even if there is no apparent connection between the substances in question.

It is crucial to inform your doctor of any current medications before starting Ciprofloxacin because it is known to interact with a variety of drugs and supplements. By doing this, you help your health care team develop a treatment plan that maximizes advantages and minimizes risks.

Medications and Ciprofloxacin

Ciprofloxacin may interact with a variety of medications, altering how your body absorbs, uses, or gets rid of it. Ciprofloxacin may interact with the following major medications:

Antacids: Drugs containing magnesium, aluminum, or calcium, which are commonly included in antacids, can significantly reduce the efficiency of Ciprofloxacin.

These minerals bind to Ciprofloxacin in the digestive tract and prevent it from being well absorbed. Antacids should be taken at least two hours before or, if necessary, six hours after your Ciprofloxacin dosage.

Blood Thinners (Warfarin): Ciprofloxacin can exacerbate the effects of warfarin and other blood thinners, increasing the risk of bleeding out of them. When you're taking a blood thinner, your doctor may need to adjust your dosage or closely monitor your blood's ability to clot.

Corticosteroids: If you are using corticosteroids (such as prednisone), ciprofloxacin significantly increases your risk of tendon damage, including tendon rupture.

This is particularly true for elderly people. Discuss this combination with your physician; they may recommend an alternative treatment plan to lower risks.

Antiarrhythmics (Medications for Heart Rhythm): Ciprofloxacin may cause serious problems with heart rhythm by interacting with amiodarone or quinidine, two medications used to treat irregular heartbeats.

Your doctor may recommend against using Ciprofloxacin if you are taking any of these drugs due to this important interaction.

Diabetes medications (insulin or oral hypoglycemics): Ciprofloxacin may alter blood sugar levels, which diabetics may find dangerous.

If you take insulin or other blood sugar-controlling drugs, your doctor will need to check your blood sugar levels frequently to prevent hypoglycemia (low blood sugar) or hyperglycemia (high blood sugar).

Seizure medications: Because ciprofloxacin lowers the seizure threshold, those with epilepsy or those using medications that control seizures, such as phenytoin or carbamazepine, may be more prone to seizures.

In order for your doctor to adjust your treatment, it is imperative that you inform them if you have ever had seizures.

Theophylline: A medication used to treat respiratory conditions like asthma, can be elevated in the blood by Ciprofloxacin. Elevated theophylline levels may cause negative side effects such nausea, vomiting, and palpitations. Your doctor might need to adjust your dosage if you are taking this medication.

Supplements and Ciprofloxacin

Supplements may also have an impact on how the body processes and absorbs Ciprofloxacin. It's important to take any of the following drugs at a different time than your Ciprofloxacin dosage in order to avoid interactions:

Calcium, Magnesium, Zinc, and Iron Supplements: In the digestive tract, supplements containing calcium, magnesium, zinc, and iron may bind to Ciprofloxacin and prevent it from entering the bloodstream.

To prevent this, take these vitamins at least two hours before taking Ciprofloxacin or six hours after.

Multivitamins: Many multivitamins contain calcium, magnesium, or iron, which can prevent Ciprofloxacin from being absorbed.

Similar to many mineral supplements, it is best to take multivitamins at alternate times than your Ciprofloxacin dosage.

Food and Ciprofloxacin Interactions

Food can also have an impact on Ciprofloxacin's effectiveness. Being mindful of what you eat while receiving treatment can help you avoid interactions that could reduce the medication's effectiveness:

Dairy products: Milk, yogurt, cheese, and other dairy products can reduce the absorption of Ciprofloxacin because calcium binds to it.

It's okay to consume dairy products, but don't take them soon before taking your Ciprofloxacin dosage. Wait at least two hours before or after taking the medicine before consuming these meals.

Caffeine: Caffeine can produce symptoms like anxiety or palpitations, and Ciprofloxacin may exacerbate these effects.

If you regularly consume tea or coffee, consider reducing your caffeine intake while taking Ciprofloxacin to avoid these side effects.

Alcohol and Ciprofloxacin

Despite the fact that alcohol and Ciprofloxacin do not directly interact, it is nonetheless recommended to limit alcohol intake when using this antibiotic. Alcohol may exacerbate the negative effects of Ciprofloxacin, which include nausea, upset stomach, and dizziness.

Additionally, excessive alcohol consumption may strain your liver and kidneys, which are already working hard to process the medicine.

To give your body the best chance of recovering, it is important to minimize your alcohol consumption throughout therapy.

Herbal Supplements and Ciprofloxacin

Even if they seem harmless, herbal supplements may interact unexpectedly with prescription medications. Ciprofloxacin and the following popular herbal supplements may not work well together:

St. John's Wort: Often used to treat depression, St. John's Wort may increase the metabolism of Ciprofloxacin, decreasing its effectiveness.

Echinacea: This is frequently used as a herbal supplement to boost immunity, but when used with certain antibiotics, including Ciprofloxacin, it increases the risk of liver damage.

To find out if any herbal supplements you may be taking are safe to take with Ciprofloxacin, speak with your doctor.

Prevention

The best defense against unpleasant experiences is to keep the lines of communication open with your health care provider.

Before starting Ciprofloxacin, make sure your doctor is aware of all the prescription, over-the-counter, and herbal medications you are currently taking. As a result, they will be more equipped to identify any potential interactions and adjust your treatment strategy accordingly.

As with other medications, supplements, or foods, follow your doctor's advice when taking Ciprofloxacin.

Proper dosage spacing can ensure that Ciprofloxacin works as intended and help prevent adverse interactions.

Ask questions and be mindful of potential interactions if you're unsure about what to take with Ciprofloxacin. Your doctor is there to support you and ensure that your therapy is as effective and safe as possible.

Chapter Seven

Overdose of Ciprofloxacin

The risk of overdose is always present when taking any medication, including Ciprofloxacin. Ciprofloxacin overdose is rare, but it can happen if you intentionally or unintentionally take more than is advised.

It is crucial to comprehend the overdose warning signs, how to respond, and the potential long-term effects. Overdosing on Ciprofloxacin can happen in a variety of ways and is typically not the consequence of deliberate misuse.

Sometimes a person may unintentionally take more medication than is advised. This could happen if the proper dosage is not known or if doses are taken too soon.

This is particularly common when family members share prescriptions, which is never a good idea.

Taking additional medications that interact with Ciprofloxacin can increase the risk of overdose by enhancing its effects or postponing its elimination from the body.

It's crucial to follow your doctor's advise regarding medication interactions in order to prevent this.

Unintentional overdose is particularly dangerous for kids who have access to medication that has been poorly maintained. To reduce this danger, Ciprofloxacin should be kept out of children's reach, just like all drugs.

Symptoms

Recognizing the signs of a Ciprofloxacin overdose is crucial for prompt action. The following are the most common symptoms, though the severity may vary:

- Dizziness or Fainting
- Confusion or Agitation

- Seizures
- Nausea and Vomiting
- Heart Palpitations or Irregular Heartbeat
- Kidney or Liver Damage

It is imperative that you get medical attention as soon as any of these symptoms emerge.

What to Do in Case of an Overdose

Overdosing is a dangerous medical condition. If you believe that you or someone else may have taken too much Ciprofloxacin, take the following steps:

Contact Emergency Services: Visit the nearest emergency room or give your local emergency number (911) a quick call.

Provide as much information as you can to the medical professionals, including the dosage, time, and existence of any other medications being used with the Ciprofloxacin.

Avoid Inducing Vomiting: Unless specifically instructed to do so by a health care provider, do not attempt to induce vomiting, even though the overdose may produce it on its own.

It may occasionally be more detrimental than helpful to induce vomiting, especially if the person is unconscious or confused.

Bring Medication Information: Try to bring your prescription information or the Ciprofloxacin box when you attend the emergency room. This will allow medical staff to assess the issue and determine the best course of action more quickly.

It might be quite distressing to be an overdose victim or witness, but it's crucial to remain calm. Make it a priority to get help as soon as you can, and follow medical professionals' instructions.

Treatment

Stabilization and recovery are the main objectives of treatment for a Ciprofloxacin overdose once the patient is admitted to the hospital. The precise treatment plan will depend on the symptoms and the extent of the overdose. Here are some examples of common interventions:

Medical professionals may occasionally administer activated charcoal to help the stomach absorb any remaining Ciprofloxacin. This could prevent the drug from further entering the bloodstream.

To help with the drug's excretion from the body, maintain hydration, and support renal function, intravenous fluids may be given. This is especially important if the overdose has resulted in diarrhea or vomiting.

The medical personnel will monitor vital signs such as heart rate, blood pressure, and respiration. In cases of severe overdose, further support, such as oxygen or medications to keep the heart functioning, may be needed.

If seizures have already occurred, anti-seizure medications may be used to protect the brain and prevent further occurrences.

Prevention

The first measures in preventing an overdose are to carefully follow the recommended dosage and to be aware of any factors that may increase the risk of overdosing. Important preventative methods include the following:

Always use Ciprofloxacin exactly as prescribed by your physician. Never change your dosage or use more medication than is recommended without first consulting your doctor.

If you struggle to remember when to take your medicine, use a medication reminder app or set alarms. By doing this, you can avoid accidentally taking too much Ciprofloxacin and ensure that you take it on time.

Ciprofloxacin should be stored safely and out of the reach of children and dogs. Use child-proof containers and store the medication in a cool, dry place as directed on the label.

Certain medications may increase the risk of overdosing because they change how Ciprofloxacin is metabolized. Always inform your doctor about any medications, herbal items, or supplements you are taking to avoid any unfavorable interactions.

Potential Long-Term Effects of Overdose

The long-term effects of a Ciprofloxacin overdose can vary depending on how severe the overdose was and how quickly it was managed. When medical care is received promptly, the majority of patients recover well.

However, in more severe cases, an overdose could result in long-term damage like:

Renal Damage: If therapy is delayed, Ciprofloxacin misuse may result in long-term renal problems.

Liver Damage: Because the liver breaks down many medications, consuming excessive amounts of them can cause liver toxicity, which can lead to long-term liver damage if treatment is not received.

Tendon Damage: An overdose may increase the risk of severe tendon injury, including tendon rupture, particularly in vulnerable individuals.

When to Seek Help

If you have any questions regarding your dosage or are concerned that you could overdose, don't be afraid to get in touch with your health care provider.

It is always better to seek clarification and ask questions than to risk making a mistake. If you see any signs that might indicate an overdose, take immediate action because it could make all the difference.

Chapter Eight

Storage

Drugs must be maintained correctly to maintain their efficacy and guarantee safety. Ciprofloxacin, like all antibiotics, needs to be carefully maintained to keep working and prevent accidental misuse.

Ciprofloxacin should be stored properly for reasons of safety and effectiveness as well as convenience. If the drug is exposed to light, heat, moisture, or other unsuitable circumstances, it may lose its effectiveness or even become hazardous to use.

By following the recommended storage guidelines, you can ensure that each dosage of Ciprofloxacin remains as reliable and effective as the first time it was administered.

Always keep Ciprofloxacin in its original container, with the label and instructions intact on it. This container is meant to protect the medication from harmful environmental factors including light and moisture.

Ciprofloxacin should be stored in a cool, dry place away from moisture and direct sunlight. Because they are often humid, toilets and kitchens are generally not the greatest places. Instead, pick a cabinet or drawer in a place that is consistently warm.

Like other medications, Ciprofloxacin has an expiration date. Using the medication after this date may reduce its effectiveness or increase the risk of adverse side effects.

Ciprofloxacin should never be used over its expiration date, and any that have gone bad should be disposed of appropriately.

Specific Storage Instructions for Different Forms

Ciprofloxacin comes in a variety of forms, each with specific storage needs.

Tablets: Ciprofloxacin pills should be stored at room temperature, often between 68°F and 77°F (20°C and 25°C). Keep them in their original bottle or blister pack, well closed, and away from moisture.

Don't keep them near windows or in your car where they can be exposed to extreme heat. After each use, make sure the bottle is securely closed to protect the pills from air and moisture.

Oral Suspension (Liquid): If you are prescribed Ciprofloxacin in liquid form, it is important that you follow the storage instructions on the label.

Most Ciprofloxacin oral suspensions should be stored in the refrigerator between 36°F and 46°F (2°C and 8°C). However, freezing the drug may reduce its effectiveness.

If refrigeration is not required, store it at room temperature, away from light and moisture. Before using, shake the bottle well to ensure the medication is mixed appropriately.

Intravenous (IV) Ciprofloxacin: If you are receiving Ciprofloxacin intravenously at home, the storage of the IV solution will depend on how it is delivered.

Certain IV solutions require refrigeration, while others can be kept at room temperature. Prior to using the solution, check for any signs of contamination, such as cloudiness or discoloration, and always follow the storage instructions provided by your doctor or pharmacist.

If the remedy seems compromised, do not use it and contact your health care provider immediately.

Eye or Ear Drops: Additionally, Ciprofloxacin eye or ear drops should be stored at room temperature, away from intense heat and light.

After each use, make sure the cap is securely secured to prevent contamination. Avoid touching the dropper tip with your hands or any other surface to preserve the solution's sterility.

Safe Disposal of Ciprofloxacin

When Ciprofloxacin is no longer needed or has expired, it should be disposed of properly to prevent accidental ingestion or harm to the environment.

Ciprofloxacin should not be flushed down the toilet or poured down the sink since it can contaminate water supplies and damage the environment.

A prescription take-back program is the safest way to dispose of Ciprofloxacin that has been unused or expired. Many pharmacies, hospitals, and local government organizations offer these services, which allow you to return any leftover medication for safe disposal.

You can dispose of Ciprofloxacin at home if there isn't a take-back program by mixing the liquid or tablets with something unattractive, such as dirt, coffee grounds, or cat litter.

The mixture should be put in a sealed plastic bag and thrown out with the household trash. This makes the medicine harder to find in the trash and less tempting to children or pets.

Keeping Ciprofloxacin Out of Reach

Always store Ciprofloxacin in a secure location to prevent accidental consumption, especially by children or pets.

The bottle or container should be kept out of children's reach, ideally in a high cabinet or locked drawer. Never leave medication on counter tops or bedside tables unattended.

Make sure that everyone in your household understands how important it is to prevent children and animals from getting their hands on medications like Ciprofloxacin. Families with young children may consider implementing extra safety precautions, such locking up prescription medications.

What to Do if Ciprofloxacin is Improperly Stored

If you find out that Ciprofloxacin has been exposed to extremes of temperature, moisture, or sunlight, you may wonder if it is still safe to use. In these situations, it is better to err on the side of caution.

If you're unsure whether a medication that has been improperly stored is still effective, contact your pharmacist. They can provide guidance on whether the medication should be replaced or whether it is still safe to take.

If your supply of Ciprofloxacin has been tainted, you must obtain a new one. Using medication that has been compromised by improper storage can reduce its effectiveness or increase the risk of negative outcomes.

Traveling with Ciprofloxacin

If you have to travel while taking Ciprofloxacin, proper storage becomes even more important.

Keep your medication in your carry-on bag whenever you travel by flight because checked luggage can be exposed to excessively high or low temperatures.

Especially important for oral suspensions that may need to be refrigerated. Consult your pharmacist about portable storage solutions like insulated bags or cold packs if you are traveling with liquid forms of Ciprofloxacin.

While a pill organizer may be helpful when traveling with medicines, be careful to save the original container so you can retrieve your prescription information and dose instructions if needed.

When traveling between time zones, talk to your health care provider about how to adjust the timing of your doses to stay on track and maintain the medication's effectiveness.

The End